CE5 Frequencies of Contact

Donald Ledesma

Published by Donald Ledesma, 2024.

CE5 FREQUENCIES OF CONTACT

First edition. April 7, 2024.

Copyright © 2024 Donald Ledesma.

ISBN: 979-8224950720

Written by Donald Ledesma.

This book is dedicated to my wife and my children, family, and to the people we have lost and missed.

Most of all I dedicate this book to my mother Glora Ann Ledesma who has always inspired me never to give up and push forward to inspire others.

Table of Contents

Forward:

    Chapter 1: Introduction to CE5

    Chapter 2: Close Encounters of the Fifth Kind (CE5) Contact: Exploring Human- Extraterrestrial Communication

    Chapter 3: Creating an Enriching CE5 Meditation Experience: Frequencies and Music Selection

    Chapter 4: Understanding Air Pressure Changes During a CE5 Experience

    Chapter 5: Contact with Extraterrestrials

    Chapter 6: Extraterrestrials Chose You

    Chapter 7: The CE5 Chosen

    Chapter 8: Meditation of a CE5

    Chapter 9: A.L.I.E.N.S- (EBE)

    Chapter 10: The Final Contact Balance

Forward

I am writing this book as an experiencer and a podcaster doing research for many years. Now as a host on a radio show called Alien Strand Podcast. With this podcast I've learned to understand many people with abductions and encounters. That's my work and I continue to help people throughout the world.

This endeavor started with me when I was 14 years old, I experienced my first UFO encounter. 42 years. One evening in 1982, as I was cutting the grass in the backyard of my mother's house, I noticed something in the sky. It was an elongated shape of a UFO crossing over the moon. Making a shadow on the moon. Ever since that day, I cannot stop pushing forward into the disclosure of ufology.

This book will help you understand the importance of a CE5 session and encounter. Throughout my lifetime, I have encountered many UFO ships as I was living in Pecos New Mexico for several years. Even some encounters I can't explain while out camping. My photos from the following morning are a bright light red light on one of my pictures that I don't remember taking that evening. Did I encounter something that I don't understand? Ever since then, I've had memories and dreams of extraterrestrial visitations in a different kind of celestial way. I believe I've had some kind of subconscious contact with these extraterrestrials. This book is a guide to open consciousness to CE5 sessions and understanding the frequencies of the human body.

Several years after watching Doctor Steven Greer explaining a CE5 session, I have learned to try these techniques myself. Do I always get a positive result? No. Frequencies do help. And puts me in a different state of mind to understand consciousness and subconsciousness. Ever since trying CE5 in my own backyard, I have noticed more strange events around my area that people were witnessing over our skies down here in South Texas. I feel that once you do a CE5 session, you are a beacon of some sort. Attracting extraterrestrials or spacecraft to your area. You are the connection. I believe anyone can do the CE 5 events.

With an open mind and understanding. You can open many possibilities to what's out there and are interstellar galactic universe. And answering the questions, are we alone? This book will help you step by step to understand what you are experiencing and what you can do to help push forward into finding the answers to those questions.

Exploring CE5 Contacting Extraterrestrial Intelligence

Chapter 1

Introduction to CE5

Close Encounters of the Fifth Kind, commonly referred to as CE5, represents a fascinating area of study that explores the potential for direct contact with extraterrestrial intelligence. Unlike traditional UFO sightings or encounters, CE5 involves proactive efforts by humans to initiate communication with beings from other worlds. This book delves into the history, methods, experiences, and implications of CE5 encounters, providing a comprehensive overview for those intrigued by the mysteries of the cosmos. Throughout my studies in ufology have been somewhat complicated but extremely forthcoming in the complete interviews of the CE contact phenomena. Their explanations of contact whether through physical or mental communication of contact. I am certain that these extraterrestrial entities are within the same realm and dimensions of human existence. Through these contact steps and mindset, the assessments are there that you are communicating with something unworldly. The cosmos and frequency changes are key to contact with your open mind looking for discoveries that only you can imagine or do. Now is the time for you to try these steps for contact.

CE5 & Dr Greer

The concept of CE5 was popularized by Dr. Steven Greer, a physician and ufologist, who founded the Center for the Study of Extraterrestrial Intelligence (CSETI). Dr. Greer proposed that through conscious intent, meditation, and peaceful interaction, humans could establish communication with advanced extraterrestrial civilizations. This idea traces its roots to ancient practices of contacting celestial beings through meditation and spiritual awareness.

Dr. Steven Greer and His Contribution to Ufology

Dr. Steven M. Greer is a prominent figure in the field of ufology, known for his research, advocacy, and efforts to promote peaceful contact with extraterrestrial civilizations. Born on June 28, 1955, in Charlotte, North Carolina, Greer's interest in UFOs and extraterrestrial intelligence began at an early age and eventually led him to pursue a career in medicine before transitioning into ufology full-time.

Early Career and Formation of CSETI

Greer completed his medical degree at the University of Virginia School of Medicine and worked as an emergency room physician. However, his passion for UFOs and the potential implications of contact with extraterrestrial beings drove him to establish the Center for the Study of Extraterrestrial Intelligence (CSETI) in 1990. CSETI aimed to conduct scientific research and facilitate peaceful communication with advanced extraterrestrial civilizations.

Disclosure Project and Government Engagement

One of Greer's most notable initiatives is the Disclosure Project, launched in 1993, which gathered testimonies from military, government, and aerospace industry insiders regarding UFOs and extraterrestrial encounters. The project aimed to urge governments to disclose classified information about UFOs and advanced propulsion technologies.

Greer's efforts to engage with government officials and institutions regarding UFO disclosure have been both praised and criticized. While some view his work as groundbreaking in raising public awareness and pushing for transparency, others criticize his methods and the perceived lack of concrete evidence to support his claims.

CE5 Protocols and Consciousness-Based Contact

Greer is best known for developing the concept of Close Encounters of the Fifth Kind (CE5), which involves proactive, conscious efforts to initiate contact with extraterrestrial civilizations. CE5 protocols emphasize meditation, consciousness expansion, group intention, and the use of light and sound signals to attract UFOs and establish communication.

Through workshops, seminars, and public events, Greer has educated and trained thousands of individuals worldwide in CE5 protocols. He believes that peaceful and respectful contact with extraterrestrial beings is possible, and that humanity can benefit from the advanced knowledge and wisdom they may offer.

Criticism and Controversies

Despite his contributions to ufology and consciousness-based contact, Greer has faced criticism and controversies throughout his career. Skeptics question the validity of his claims, citing the lack of scientific evidence and verifiable data supporting his assertions about UFOs and extraterrestrial encounters.

Additionally, Greer's approach to UFO disclosure and his public persona have sparked debates within the ufology community. Some researchers and enthusiasts view him as a visionary and pioneer, while others are skeptical of his methods and the commercialization of UFO-related events and materials.

Legacy and Impact

Regardless of the controversies surrounding him, Dr. Steven Greer has left a significant impact on ufology and the exploration of extraterrestrial intelligence. His work has inspired countless individuals to delve into the mysteries of UFOs, consciousness, and the potential for peaceful contact with other civilizations. Whether regarded as a visionary or a controversial figure, Greer's influence on the discourse surrounding UFOs and extraterrestrial phenomena is undeniable.

Methods of CE5

CE5 methods often involve gathering a group of individuals who share a common intention to make contact with extraterrestrial beings. These gatherings may take place in remote locations with minimal light pollution, allowing for optimal sky watching. Participants engage in meditation, visualization exercises, and use technologies such as lasers or light signals to attract the attention of potential extraterrestrial craft.

Experiences of CE5 Practitioners

Numerous individuals and groups worldwide have reported compelling experiences during CE5 sessions. These experiences range from sightings of unusual aerial phenomena to alleged direct communication with non-human entities. Accounts often describe feelings of interconnectedness, profound spiritual insights, and a sense of awe and wonder at the vastness of the universe and humanity's place within it.

During a Close Encounter of the Fifth Kind (CE5), individuals often report a range of experiences that can be profound, transformative, and sometimes challenging to describe within conventional frameworks. These experiences can vary widely depending on the individual's level of consciousness, sensitivity, and the nature of the contact. Here are some common themes of what people experience during a CE5 event.

1. **Visual Sightings:** Many participants in CE5 sessions report seeing unusual lights, orbs, or structured craft in the sky. These sightings are often described as moving in ways that defy known laws of physics, such as sudden changes in direction or acceleration. Witnesses may observe these phenomena alone or in groups, enhancing the credibility of their accounts. Photos are usually reliable during the CE5 event giving you more concrete evidence to prove the event happened.

2. **Telepathic Communication:** Some individuals claim to have received telepathic messages or impressions during CE5 contact. These communications are often described as non-verbal, with information being transmitted directly to the recipient's consciousness. Messages may convey insights into the nature of reality, consciousness, spirituality, and the purpose of human existence.

3. **Energetic Sensations:** CE5 participants frequently report feeling heightened states of energy or vibration during contact experiences. These sensations may manifest as tingling, warmth, or a sense of

expansion in consciousness. Some describe feeling a deep sense of peace, love, and interconnectedness with all life forms.

4. **Emotional Responses:** Contact with extraterrestrial beings or advanced consciousness can evoke a wide range of emotions, including awe, wonder, joy, fear, and confusion. Participants may undergo profound emotional shifts as they confront the reality of interstellar communication and the implications for humanity's place in the universe. The memory of communication will stay with you during consciousness and will show up in dreams within the subconscious realm.

5. **Synchronicities and Insights:** CE5 experiences are often accompanied by synchronicities or meaningful coincidences in participants' lives. These synchronicities may include dreams, visions, or encounters with symbolic animals or symbols related to extraterrestrial communication. Participants may also gain new insights into their life purpose, spiritual path, and connection to the cosmos. Some DeJa'Vu moments may intensify because of the awakening moment of your consciousness. You will have a clearer mind of thinking as moment of contact intensified by the spiritual emitting invisible photon transfer to your body as it happened before your eyes.

6. **Healing and Transformation:** CE5 experiences are frequently described as healing and transformative, leading to personal growth, spiritual awakening, and expanded awareness. Participants may undergo profound shifts in consciousness, beliefs, and worldviews as they integrate their contact experiences into their lives. Some healing of prior physical issues may occur after a CE5 event within weeks after the positive charge of frequencies to the human body. A better sense of wellness will be present as you start to notice during the weeks after the event. Overall, what people experience during a CE5 can be deeply subjective and multifaceted, encompassing a spectrum of sensory, emotional, and spiritual phenomena. These experiences challenge our

understanding of reality and invite us to explore the mysteries of human-extraterrestrial communication with open minds and hearts.

Challenges and Skepticism

While many proponents of CE5 attest to transformative experiences and meaningful encounters, the field also faces challenges and skepticism. Critics argue that anecdotal evidence lacks scientific rigor and that many reported sightings can be attributed to natural or man-made phenomena. The lack of verifiable, reproducible evidence remains a significant hurdle in gaining wider acceptance of CE5 within the scientific community. Data is usually what most skeptics are looking for during or after a CE5 contact. Be sure to always have a person from the outside collecting data as well for the proof of contact that you will want yourself and to show skeptics out there, that contact through these procedures does exist in our human realm.

The Science of Consciousness and Interstellar Communication

Advances in the study of consciousness and quantum physics have sparked renewed interest in the possibility of interstellar communication. Theoretical frameworks such as quantum entanglement and non-local consciousness propose mechanisms through which information could be exchanged across vast distances, potentially facilitating communication with extraterrestrial intelligences. Consciousness is to believe to travel through the universe as we can alter the connectivity to extraterrestrials and start to communicate with the entities. This will be the beginning of "The Calling" almost like a phone call with an unlimited signal to many civilizations in the universe. We will just be waiting for the right frequency to answer the call for "Hello".

Ethical Considerations and Intentions

CE5 practitioners often emphasize the importance of approaching contact experiences with respect, peaceful intentions, and ethical considerations. Advocates promote non-violence, empathy, and cooperation as guiding principles in human-extraterrestrial interactions. These values reflect a broader shift toward a more compassionate and interconnected worldview. The right mindset for communication from a CE5 event is key to the ethical standards of friendliness, compassion, and empathy to the visitation.

Implications for Humanity

The exploration of CE5 and the quest for contact with extraterrestrial civilizations raise profound questions about humanity's place in the cosmos and the nature of consciousness. If successful, such contact could revolutionize our understanding of reality, spirituality, and the potential for peaceful coexistence on a planetary and cosmic scale.

The Future of CE5 Research

As interest in CE5 continues to grow, researchers and enthusiasts explore new avenues for investigation. Advances in technology, such as improved sky monitoring devices and data analysis techniques, offer opportunities to gather more robust evidence of anomalous phenomena and potential contact events. Collaboration between scientists, experiencers, and citizen scientists holds promise for advancing our knowledge in this field. Ufologists all over the world can collect the data necessary for the science of true contact. Research is key to the total involvement with CE5 exploration throughout the world. Video proof and photographic proof are necessary to complete this mission of disclosure.

CE5 represents a frontier of exploration that bridges science, spirituality, and the unknown. Whether viewed as a quest for cosmic companionship, a spiritual journey, or a scientific endeavor, CE5 invites us to contemplate the mysteries of existence and our place in a vast and diverse universe. As we continue to explore the realms beyond our planet, may we approach these encounters with curiosity, humility, and a sense of wonder that transcends boundaries and unites us in our shared cosmic journey.

Chapter 2

Close Encounters of the Fifth Kind (CE5) Contact: Exploring Human-Extraterrestrial Communication

Close Encounters of the Fifth Kind (CE5) refers to a concept in ufology and extraterrestrial studies that involves direct, intentional communication between humans and extraterrestrial beings or their spacecraft. Unlike other close encounters that are passive or involuntary, CE5 contact is proactive, initiated by humans seeking to establish peaceful and meaningful interactions with non-human intelligences. This report delves into the background, methods, experiences, and implications of CE5 contact.

The term "Close Encounters of the Fifth Kind" was coined by Dr. Steven M. Greer, a prominent figure in the field of ufology and the founder of the Center for the Study of Extraterrestrial Intelligence (CSETI). CE5 is based on the classification system developed by Dr. J. Allen Hynek, who categorized UFO sightings into different levels, with CE1 to CE4 describing varying degrees of visual contact with unidentified flying objects (UFOs) or aliens.CE5 represents the highest level of interaction, where humans actively engage in communication with extraterrestrial entities. This communication is not limited to traditional methods like radio signals but can also include telepathic, meditative, or consciousness-based exchanges.

Close Encounter Levels Explained:

Dr. J. Allen Hynek's classification system for close encounters with UFOs, often referred to as the Hynek Scale, categorizes sightings into different levels based on the nature of the encounter and the degree of interaction with the unidentified flying object (UFO) or extraterrestrial phenomena. The scale includes three primary categories: Close Encounters of the First Kind (CE1), Close Encounters of the Second Kind (CE2), and Close Encounters of the Third Kind (CE3). Let's explore each category in detail:

1. Close Encounters of the First Kind (CE1):

CE1 encounters involve visual sightings of UFOs at close range, typically within 500 feet. These sightings often include detailed observations of the object's appearance, shape, size, color, and movement patterns. Witnesses may describe seeing structured craft, unusual lights, or aerial phenomena that cannot be readily identified as conventional aircraft or natural phenomena.

One key characteristic of CE1 encounters is the direct visual contact between the observer and the UFO. Witnesses may report seeing the object hover, maneuver, or perform rapid movements that defy known aviation capabilities. Physical effects such as electromagnetic interference, radiation burns, or disruptions to electronic devices may also occur during CE1 encounters.

While CE1 encounters provide valuable visual data and firsthand observations, they are often limited in terms of additional sensory experiences or interactions with the UFO or its occupants. These sightings contribute to the documentation and analysis of UFO phenomena but do not involve direct communication or physical contact with extraterrestrial beings.

2. Close Encounters of the Second Kind (CE2):

CE2 encounters go beyond visual sightings and involve physical effects or environmental anomalies associated with the presence of a UFO. These effects can include electromagnetic interference (EMI) affecting vehicles or electronic devices, disruptions to communication systems, power outages, or changes in the surrounding environment such as vegetation damage or soil disturbances.

In CE2 encounters, witnesses may experience tangible evidence of the UFO's interaction with the environment, leading to measurable effects that can be documented and investigated. These effects contribute to the credibility and scientific study of UFO phenomena, as they provide empirical data beyond subjective eyewitness testimony.

However, CE2 encounters still lack direct communication or contact with extraterrestrial beings. The focus remains on observing and documenting the physical effects and anomalies associated with UFO presence, often requiring interdisciplinary investigation involving scientists, engineers, and researchers.

3. Close Encounters of the Third Kind (CE3):

CE3 encounters represent the highest level of interaction and involvement with UFOs, involving direct contact or communication with extraterrestrial beings or occupants of the UFO. These encounters may include visual sightings of humanoid or non-human entities associated with the UFO, encounters with landed or hovering craft, or interactions with beings through telepathic communication.

One of the defining features of CE3 encounters is the presence of conscious entities or intelligence associated with the UFO, leading to meaningful exchanges or interactions with witnesses. These encounters can be transformative, evoking emotional, spiritual, or philosophical responses as individuals confront the reality of contact with non-human intelligence.CE3 encounters often involve multiple witnesses who corroborate each other's testimonies, providing additional credibility and validation to the experience. These encounters may also include physical evidence such as trace markings, physiological effects on witnesses, or recorded data that supports the reality of the encounter. In summary, Dr. J. Allen Hynek's classification system of close encounters provides a structured framework for understanding and categorizing UFO sightings based on the level of interaction, physical effects, and presence of conscious entities. Each category contributes to the study of UFO phenomena and the exploration of potential extraterrestrial visitation, offering valuable insights into the nature of human-extraterrestrial encounters.

Dr. J. Allen Hynek's close encounter classification system includes additional categories beyond CE1, CE2, and CE3, specifically Close Encounters of the Fourth Kind (CE4) and Close Encounters of the Fifth Kind (CE5). These categories extend the spectrum of encounters to include more immersive and transformative experiences with UFOs and extraterrestrial phenomena. Let's explore each category in detail:

Close Encounters of the Fourth Kind (CE4):

CE4 encounters involve direct contact or interaction with extraterrestrial beings or entities, often characterized by abduction experiences, onboard UFO encounters, or close encounters with humanoid or non-human intelligence associated with the UFO. Unlike CE3 encounters, which primarily focus on visual or telepathic communication, CE4 encounters involve physical contact or abduction scenarios.

Abduction experiences reported in CE4 encounters often include memories of being taken aboard a UFO, undergoing medical examinations, or engaging in communication with non-human entities. Witnesses may describe missing time, unusual markings or implants on their bodies, and psychological effects such as post-traumatic stress disorder (PTSD) or altered perceptions of reality.CE4 encounters are highly controversial and subject to intense scrutiny within the ufology community and scientific circles. Skeptics often attribute abduction experiences to psychological factors, sleep disorders, or false memories, while proponents argue for the reality of these encounters based on consistent patterns and testimonies from multiple witnesses.

Close Encounters of the Fifth Kind (CE5):

CE5 encounters represent the highest level of proactive, intentional communication and contact with extraterrestrial civilizations, often initiated by humans using consciousness-based methods and protocols. These encounters involve telepathic communication, visual sightings of UFOs in response to human signals, and cooperative interactions between humans and non-human intelligence.

Key elements of CE5 encounters include:

◇ Consciousness-Based Communication: CE5 protocols emphasize meditation, group intention, and telepathic communication as means to establish contact with extraterrestrial beings. Participants may receive telepathic messages, insights, or downloads of information during these interactions.

- Light and Sound Signals: CE5 practitioners often use light signals such as flashlights, laser pointers, or sound signals to attract the attention of UFOs or communicate with them. These signals are believed to convey messages or invitations for contact.

- Group Dynamics: CE5 encounters are frequently conducted in groups, where collective intention and coherence amplify the potential for contact. Participants may report synchronized sightings, shared telepathic experiences, or heightened states of consciousness during group sessions.

- Documentation and Data Collection: CE5 experiences are documented through photographs, videos, audio recordings, and written reports. This data collection helps validate and analyze experiences for research and scientific inquiry.

CE5 encounters challenge conventional paradigms of human-extraterrestrial communication, emphasizing the role of consciousness, intention, and mutual respect in establishing peaceful and meaningful contact with advanced civilizations. While skeptics question the verifiability of CE5 experiences, proponents advocate for continued exploration and research into consciousness-based contact as a pathway to interstellar diplomacy and understanding.

Methods of CE5 Contact-(Steps)

There are several methods and protocols developed by Dr. Steven Greer and CSETI for initiating and maintaining CE5 contact. These methods are designed to facilitate a peaceful and respectful interaction with potential extraterrestrial civilizations. Some of the key methods include:

1. Meditation and Consciousness Expansion: CE5 practitioners often use meditation and consciousness-expanding techniques to raise their awareness and attune themselves to higher states of consciousness. This is believed to facilitate telepathic communication with extraterrestrial beings. Sit in silence for 20 to 30 min to lock in your state of consciousness. Meditate giving yourself calm thoughts and passive intent for "The Calling". Clear your mind. Make sure no children or pet distractions are not present during the exercise.

2. Group Intent and Visualization: CE5 contact is often conducted in groups, where participants collectively focus their intention on making contact with benevolent extraterrestrial beings. Visualization exercises are used to create a welcoming and inviting atmosphere for contact.

3. Use of Light and Sound Signals: Some CE5 practitioners use light signals such as flashlights or laser pointers to attract the attention of UFOs or communicate with them. These signals are often accompanied by specific sequences or patterns to convey messages.

4. Remote Viewing and Consciousness Projection: Advanced CE5 techniques may involve remote viewing or consciousness projection, where individuals project their awareness beyond the physical body to connect with extraterrestrial consciousness or spacecraft.

5. Data Collection and Documentation: CE5 experiences are often documented through photographs, videos, audio recordings, and written reports. This data collection helps validate and analyze experiences for scientific and research purposes.

Experiences and Observations

Many individuals and groups around the world claim to have had CE5 experiences, ranging from visual sightings of UFOs to telepathic communication with extraterrestrial beings. These experiences are often profound and transformative, leading to shifts in consciousness, beliefs, and worldviews. Some common themes and observations from CE5 experiences include:

1. *Visual Sightings: Participants* in CE5 sessions often report seeing unusual lights, orbs, or structured craft in the sky. These sightings are sometimes accompanied by anomalous movements or behaviors that defy conventional explanations.

2. *Telepathic Communication*: Some individuals claim to have received telepathic messages or downloads of information during CE5 contact. These messages may convey insights into the nature of reality, consciousness, and the universe.

3. *Healing and Transformation*: CE5 experiences are often described as healing and transformative, leading to personal growth, spiritual awakening, and expanded awareness. Participants report feelings of interconnectedness, love, and unity with all life forms.

4. *Synchronicities and Coincidences*: CE5 practitioners often report synchronicities or meaningful coincidences surrounding their contact experiences. These synchronicities may include dreams, visions, or encounters with symbolic animals or symbols related to extraterrestrial communication.

5. *Scientific Interest*: While CE5 experiences are primarily subjective and anecdotal, there is growing scientific interest in studying these phenomena. Researchers are exploring the use of technology, such as radar and infrared cameras, to capture objective data during CE5 encounters.

Implications of CE5 Contact

The concept of CE5 contact carries significant implications for humanity's understanding of consciousness, extraterrestrial life, and the nature of reality. Some of these implications include:

1. Expanded Consciousness: CE5 experiences challenge conventional notions of consciousness and highlight the interconnectedness of all beings in the cosmos. They suggest that consciousness is not limited to the human brain but exists as a universal phenomenon that can be accessed and shared across dimensions.

2. Interstellar Diplomacy: CE5 contact advocates argue that peaceful and respectful communication with extraterrestrial civilizations can pave the way for interstellar diplomacy and cooperation. They believe that humanity has much to learn from advanced ET civilizations in areas such as technology, spirituality, and sustainability.

3. Paradigm Shift: CE5 experiences have the potential to catalyze a paradigm shift in human consciousness, moving towards a more holistic and inclusive worldview. They invite us to reconsider our place in the universe and our relationship with other intelligent beings.

4. Ethical Considerations: CE5 contact raises ethical considerations regarding our responsibility as stewards of Earth and representatives of humanity in interstellar interactions. Advocates emphasize the importance of approaching contact with integrity, humility, and goodwill.

Close Encounters of the Fifth Kind (CE5) contact represents a fascinating and evolving frontier in the study of human-extraterrestrial communication. While the phenomena associated with CE5 experiences are often subjective and challenging to quantify scientifically, they hold profound implications for our understanding of consciousness, interstellar relations, and the nature of reality. Continued research, exploration, and dialogue are essential for navigating this complex and transformative aspect of human experience.

Chapter 3

Creating an Enriching CE5 Meditation Experience: Frequencies and Music Selection

Close Encounters of the Fifth Kind (CE5) meditations are designed to facilitate peaceful and intentional contact with extraterrestrial intelligences through heightened states of consciousness. Central to the effectiveness of CE5 meditations are the frequencies and music used, which play a crucial role in enhancing relaxation, deepening meditation, and fostering a connection with the cosmos. In this comprehensive guide, we'll explore the science behind frequencies, the role of music in meditation, and specific recommendations for frequencies and music tracks to use during CE5 meditation sessions.

Understanding Frequencies

Frequencies, in the context of meditation and consciousness exploration, refer to the vibrations or oscillations of energy that can influence brainwave patterns, mental states, and emotional experiences. Different frequency ranges are associated with specific states of consciousness, such as relaxation, focus, creativity, and transcendence. Here are some key frequency ranges and their corresponding effects:

- Delta Waves (0.5-4 Hz): Delta waves are associated with deep sleep, unconsciousness, and regeneration. Using delta wave frequencies can promote deep relaxation, restorative sleep, and access to the subconscious mind during meditation.
- Theta Waves (4-8 Hz): Theta waves are linked to deep meditation, creativity, intuition, and dream states. Theta frequencies can enhance visualization, inner exploration, and accessing deeper layers of consciousness during meditation.

- Alpha Waves (8-14 Hz): Alpha waves are associated with relaxed wakefulness, calmness, and a meditative state of mind. Alpha frequencies can promote stress reduction, mental clarity, and a sense of inner peace during meditation.

- Beta Waves (14-30 Hz): Beta waves are present during active waking states, focused attention, and cognitive tasks. Using beta wave frequencies can enhance concentration, alertness, and cognitive processing during meditation.

- Gamma Waves (30-100 Hz): Gamma waves are associated with heightened states of awareness, peak performance, and spiritual experiences. Gamma frequencies can promote expanded consciousness, enhanced perception, and states of transcendence during meditation.

The Role of Music in Meditation

Music is a powerful tool in meditation, as it can evoke emotions, create a conducive environment, and guide the mind into deeper states of relaxation and concentration. When selecting music for CE5 meditations, consider the following elements:

- *Instrumentation*: Choose music that includes calming instruments such as flutes, strings, bells, or nature sounds like flowing water or birdsong. Avoid music with harsh or jarring sounds that can disrupt the meditative experience.

- *Tempo and Rhythm*: Opt for music with a slow tempo and gentle rhythm to promote relaxation and a sense of flow. Avoid music with rapid beats or fluctuations that can be distracting during meditation.

- Melody and Harmony: Look for music with soothing melodies and harmonies that resonate with a sense of peace, serenity, and expansiveness. Avoid music with dissonant or discordant elements that can create tension or discomfort.

- Duration: Choose music tracks that are long enough to support an extended meditation session, typically ranging from 30 minutes to an hour or more. Avoid abrupt transitions or interruptions in the music that can disrupt the meditative flow.

Recommended Frequencies and Music for CE5 Meditation

Now, let's explore specific frequencies and music tracks that are well-suited for enhancing the CE5 meditation experience:

- Delta Wave Frequencies:
- Frequency Range: 0.5-4 Hz
- Effects: Deep relaxation, sleep, subconscious exploration

Recommended Music

- "Delta Sleep Music" by Binaural Beats Sleep
- "Deep Delta Meditation" by Brainwave Power Music
- "Disclosure/CE5/ by Mirmir's Well- (YouTube)
- "1 Hour of CE5 Inspired Meditation Music" by Meditation4u (YouTube)
- "CE5 Group Contact Meditation 15 Minutes Guided by: Radiant Heart Meditation

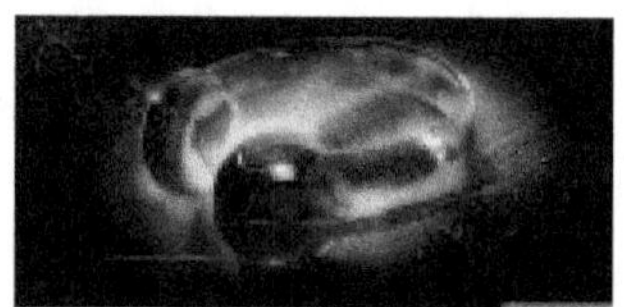

Theta Wave Frequencies

- Frequency Range: 4-8 Hz
- Effects: Deep meditation, creativity, intuition, inner exploration

Recommended Music:

- "Theta Meditation Music" by ZenLifeRelax
- "Theta Waves Meditation Music" by NuMeditationMusic

Alpha Wave Frequencies:

- Frequency Range: 8-14 Hz
- Effects: Relaxation, calmness, mental clarity, stress reduction

Recommended Music:

- "Alpha Meditation Music" by Yellow Brick Cinema
- "Calm Alpha Waves" by Power Thoughts Meditation Club

Beta Wave Frequencies

- Frequency Range: 14-30 Hz
- Effects: Focus, alertness, cognitive processing

Recommended Music:

- "Beta Focus Music" by Relaxing Music Therapy
- "Binaural Beats Beta Waves" by Meditative Mind

Gamma Wave Frequencies

- Frequency Range: 30-100 Hz
- Effects: Heightened awareness, peak performance, expanded consciousness

Recommended Music:

- "Gamma Wave Music" by Binaural Beats Meditation

(Brainwave Power Music)
- "Gamma Waves Meditation Music" by ZenLifeRelax

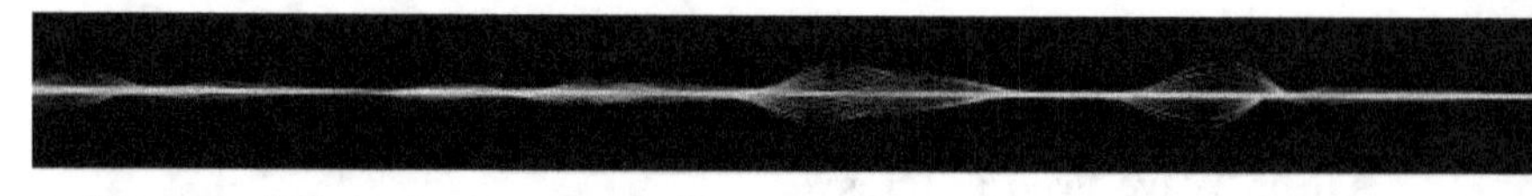

Additional Tips for CE5 Meditation

1._Create a Sacred Space_: Set up a quiet and comfortable space for meditation, free from distractions and interruptions. Use candles, incense, crystals, or sacred objects to enhance the ambiance and create a sacred atmosphere.

2. _Set Intentions_: Before starting the meditation, set clear intentions for the CE5 contact experience. Focus on openness, receptivity, and peaceful communication with extraterrestrial intelligences.

3. _Practice Breathing Techniques_: Incorporate deep breathing techniques such as diaphragmatic breathing, alternate nostril breathing, or box breathing to enhance relaxation, oxygenation, and mindfulness during meditation.

4._Visualize Light and Connection_: During meditation, visualize a sphere of light surrounding you and expanding outward, creating a field of connection and communication with the cosmos. Imagine sending out signals of peace, love, and harmony to attract benevolent extraterrestrial beings.

5._Journaling and Integration_: After meditation, take time to journal your experiences, insights, and impressions. Reflect on any messages, symbols, or sensations you may have received during the CE5 contact. Integrate these experiences into your daily life and spiritual practice.

By incorporating the right frequencies, music, and mindful practices into your CE5 meditation sessions, you can enhance your ability to connect with extraterrestrial intelligences, explore expanded states of consciousness, and foster a deeper sense of unity with the cosmos. Experiment with different music tracks, frequencies, and meditation techniques to find what resonates best with your personal journey of CE5 exploration and contact. Music is key.

Chapter 4

Understanding Air Pressure Changes During a CE5 Experience

Close Encounters of the Fifth Kind (CE5) experiences often involve a range of sensory phenomena, including changes in air pressure. These fluctuations in atmospheric pressure can be subtle or dramatic, depending on the intensity of the contact and the environmental conditions. In this comprehensive exploration, we will delve into the science behind air pressure changes during CE5 encounters, the potential causes and mechanisms involved, and the significance of these phenomena in the context of human-extraterrestrial communication.

The Science of Air Pressure

Air pressure, also known as atmospheric pressure, refers to the force exerted by the weight of air molecules in the Earth's atmosphere. It is measured in units such as millibars (mb) or inches of mercury (inHg) and varies with altitude, weather conditions, and geographical location. Normal atmospheric pressure at sea level is around 1013.25 millibars (29.92 inches of mercury).

Changes in air pressure can occur due to various factors, including weather systems (such as high and low-pressure systems), temperature fluctuations, altitude variations, and atmospheric disturbances. These changes can impact human perception, physical sensations, and environmental conditions.

Air Pressure Changes in CE5 Experiences

During CE5 experiences, witnesses often report noticeable changes in air pressure that coincide with the presence of unidentified flying objects (UFOs), extraterrestrial beings, or energetic phenomena. These air pressure fluctuations may manifest as:

- Pressure Build-Up: Some witnesses describe a sense of pressure or heaviness in the air, as if the atmosphere is denser or more compressed than usual. This sensation can be accompanied by a feeling of being "pressurized" or enclosed within a distinct field of energy.

- Pressure Drop: Conversely, others report a sudden drop in air pressure, akin to a vacuum or expansion of space around them. This sensation can create a feeling of lightness, openness, or expansiveness, as if the environment has become more spacious and permeable.

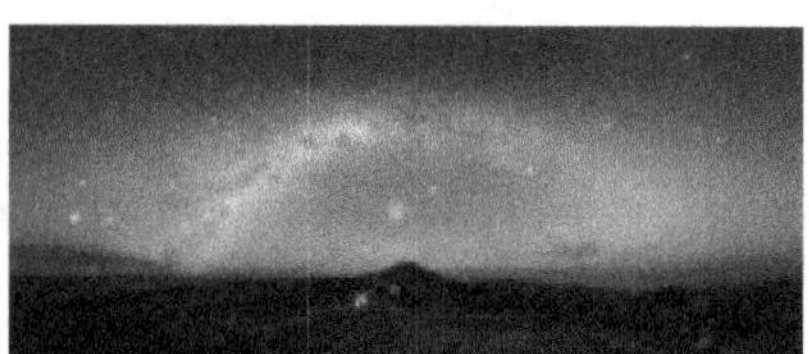

Pulsating or Vibrating Pressure: In some cases, witnesses describe a pulsating or vibrating sensation in the air, as if the atmosphere is oscillating or resonating with energetic frequencies. This pulsation can be rhythmic or irregular, leading to a dynamic perception of the surrounding space.

- Localized Pressure Changes: Air pressure fluctuations during CE5 encounters may also be localized, affecting specific areas or zones within the immediate vicinity of the witnesses. This localized effect can be spatially confined yet perceptually distinct, highlighting the dynamic nature of the atmospheric changes.

Potential Causes and Mechanisms

The causes and mechanisms underlying air pressure changes during CE5 experiences are not fully understood and may involve a combination of physical, energetic, and consciousness-based factors. Some potential explanations include:

Energetic Interactions: It is theorized that the presence of advanced extraterrestrial technologies or energetic fields associated with UFOs can influence the surrounding atmosphere, leading to changes in air pressure. These energetic interactions may result from electromagnetic effects, gravitational anomalies, or unknown energy phenomena.

Consciousness Effects: Human consciousness and intentionality may play a role in modulating air pressure during CE5 encounters. It is postulated that focused mental states, collective intentionality within a group, and telepathic communication with extraterrestrial intelligences can influence the energetic dynamics of the environment, including air pressure fluctuations.

Electromagnetic Fields: UFOs and advanced spacecraft are often associated with strong electromagnetic fields that can affect atmospheric conditions. These electromagnetic effects may contribute to changes in air pressure, electromagnetic interference with electronic devices, and anomalous sensory experiences reported by witnesses.

Dimensional Shifts: Some theories propose that CE5 experiences involve shifts in dimensional or vibrational frequencies, leading to alterations in the perception of space, time, and physical properties such as air pressure. These shifts may occur temporarily during contact events, resulting in perceptual anomalies and sensory transformations.

Significance and Interpretation

The significance of air pressure changes during CE5 experiences extends beyond the physical phenomena and raises profound questions about the nature of reality, consciousness, and interstellar communication. Here are some interpretations and considerations:

- *Energetic Resonance*: Air pressure fluctuations may reflect the energetic resonance between human consciousness, extraterrestrial intelligences, and the surrounding environment. These fluctuations could signify a harmonious alignment or interaction of energetic fields, transcending conventional boundaries of space and matter.

- *Symbolic Manifestations*: Air pressure changes may serve as symbolic manifestations of energetic shifts, consciousness expansion, and transformative experiences during CE5 encounters. They can symbolize the merging of inner and outer realities, the dissolution of perceived limitations, and the emergence of new paradigms of understanding.

- *Confirmation of Contact*: For witnesses and participants in CE5 sessions, air pressure changes can serve as validation or confirmation of contact with non-human intelligence. These changes, when experienced collectively within a group, can reinforce the reality of the contact event and strengthen the sense of interconnectedness and shared experience.

- *Quantum Perspective*: From a quantum perspective, air pressure fluctuations during CE5 encounters may be viewed as quantum fluctuations or perturbations in the fabric of spacetime. These fluctuations could be indicative of quantum entanglement, non-locality, and the interplay of consciousness with quantum phenomena.

Practical Considerations and Experiences

In practical terms, witnesses and participants in CE5 experiences can take several steps to observe, document, and interpret air pressure changes:

Measurement Devices: Utilize portable air pressure measurement devices, such as barometers or digital pressure sensors, to monitor and record air pressure changes during CE5 encounters. Document the time, location, and intensity of the fluctuations for further analysis. Use EMF (Electronic Magnetic Field) application on a cell phone or an actual handheld meter to measure fields during the session.

Observational Techniques: Encourage witnesses to observe and describe their sensory experiences related to air pressure changes, including any physical sensations, perceptual shifts, or environmental anomalies. Create a structured framework for reporting and documenting these observations.

Group Dynamics: Explore the role of collective consciousness and group dynamics in influencing air pressure fluctuations during CE5 sessions. Foster open communication, shared intentionality, and mutual support within the group to enhance the coherence and resonance of the contact experience.

Integration and Reflection: After the CE5 experience, engage in reflective practices, group discussions, and integrative exercises to process the air pressure changes and their implications. Encourage participants to share their insights, interpretations, and personal transformations resulting from the contact event.

Case Studies and Anecdotal Evidence

While scientific research on air pressure changes during CE5 experiences is limited, there are numerous anecdotal accounts and case studies that highlight the reality and significance of these phenomena. Here are a few examples:

Case Study 1: A group of CE5 practitioners in a remote location reported a sudden drop in air pressure during a meditation session focused on contact with extraterrestrial beings. Witnesses described feeling a sense of expansion and lightness, as if the atmosphere had become more spacious and receptive to higher frequencies.

Case Study 2: An individual conducting solo CE5 practices in a natural setting observed a localized pressure build-up around a specific area where a UFO was sighted. The witness noted a palpable sense of energy and presence, accompanied by visual anomalies and electromagnetic disturbances.

Case Study 3: During a CE5 event involving a large group of participants, air pressure changes were documented using portable barometers and pressure sensors. Fluctuations in air pressure.

Chapter 5

Contact with Extraterrestrials

Signals Sent to Space for Extraterrestrial Contact Exploring Communication Initiatives and Challenges

The quest for contact with extraterrestrial civilizations has led to numerous initiatives involving the transmission of signals into space. These signals, known as interstellar messages or METI (Messaging Extraterrestrial Intelligence), are designed to communicate information about humanity, Earth, and our desire for peaceful contact with intelligent beings beyond our planet. In this comprehensive exploration, we will delve into the history of interstellar messaging, the scientific and ethical considerations, the challenges and opportunities of signal transmission, and the potential implications for humanity's future.

Historical Overview of Interstellar Messaging

The concept of sending deliberate signals into space to communicate with extraterrestrial intelligences dates back to the mid-20th century. Here are key milestones in the history of interstellar messaging.

1959- Project Ozma - In a pioneering effort, astronomer Frank Drake conducted Project Ozma, using a radio telescope to search for signals from potential extraterrestrial civilizations. Although no signals were detected, this project laid the groundwork for future SETI (Search for Extraterrestrial Intelligence) initiatives.

1974- Arecibo Message - Scientists at the Arecibo Observatory in Puerto Rico transmitted the Arecibo Message, a binary-coded pictorial message containing information about Earth's location, human DNA, and key scientific concepts. This transmission was directed towards the globular star cluster M13, located approximately 25,000 light-years away.

1999-Cosmic Call - The Cosmic Call project, led by radio astronomer Alexander Zaitsev, sent interstellar messages containing music, greetings, and mathematical sequences to selected star systems. These messages aimed to showcase human culture, intelligence, and curiosity about the cosmos.

2008- Hello From Earth - The Hello from Earth initiative invited people worldwide to submit messages to be transmitted to Gliese 581d, a potentially habitable exoplanet. This collaborative effort symbolized global interest in interstellar communication and the search for extraterrestrial life.

2020-Sónar Calling GJ273b - The Sonar Calling GJ273b project, organized by METI International, transmitted digital messages and music compositions to the exoplanet GJ 273b (also known as Luyten's Star b) using radio telescopes. The messages aimed to convey peaceful intentions and cultural diversity to potential extraterrestrial recipients.

Scientific and Ethical Considerations

The transmission of signals into space for extraterrestrial contact raises important scientific, ethical, and philosophical questions. Here are key considerations:

Scientific Goals: Interstellar messaging initiatives aim to expand our understanding of the cosmos, explore the potential for life beyond Earth, and engage in cross-cultural communication with hypothetical extraterrestrial civilizations. These initiatives contribute to astrobiology, astronomy, and the search for intelligent life in the universe.

Ethical Framework: Ethical considerations in interstellar messaging include issues of consent, risk assessment, potential consequences, and the impact on Earth's biosphere. Critics argue that sending signals into space could attract unwanted attention from advanced civilizations or pose existential risks if misinterpreted or intercepted by hostile entities.

Cosmic Zoo Hypothesis: The "cosmic zoo" hypothesis suggests that extraterrestrial civilizations may be observing Earth discreetly without revealing their presence. Some proponents of this hypothesis caution against active signal transmission, fearing it could disrupt the balance of interstellar relations or invite unintended consequences.

Precautionary Principle: The precautionary principle advocates for careful deliberation and risk assessment before engaging in activities that could have irreversible or unknown consequences. This principle is relevant to interstellar messaging efforts, prompting researchers to consider the potential impacts and unintended outcomes of signal transmission.

Challenges and Opportunities of Signal Transmission

The transmission of signals into space for extraterrestrial contact faces several challenges and opportunities. Our current technology limits the range and scope of interstellar messaging, as radio signals weaken over vast distances and encounter interference from cosmic noise. Advanced communication technologies, such as directed energy beams or optical lasers, may offer more efficient methods of signal transmission in the future. Selecting appropriate target stars or exoplanets for interstellar messages requires careful consideration of factors such as distance, stellar characteristics, planetary conditions, and the likelihood of hosting intelligent life. Target selection criteria vary among different messaging initiatives based on scientific, cultural, and strategic priorities.

As we are crafting the content of interstellar messages involves decisions about linguistic representation, symbolic encoding, scientific information, cultural references, and the intended message recipients. Messages may include mathematical sequences, scientific diagrams, cultural artifacts, music, greetings, and symbolic imagery. Our interpretation of interstellar messages by hypothetical extraterrestrial recipients poses challenges due to linguistic and cultural differences, symbolic ambiguity, and the limitations of mutual understanding. Efforts to develop universal languages, mathematical symbols, and pictorial representations aim to enhance the comprehensibility of messages. Interstellar messaging initiatives require long-term funding, international collaboration, public engagement, and ongoing scientific evaluation. Ensuring the sustainability and continuity of these initiatives is essential for advancing our knowledge of the cosmos and maintaining global interest in space exploration and communication.

Initiatives and Protocols for Interstellar Messaging

Several organizations and initiatives are actively involved in interstellar messaging and communication protocols:

METI International: METI International (Messaging Extraterrestrial Intelligence) advocates for proactive signal transmission and interstellar messaging as a means of initiating peaceful contact with extraterrestrial civilizations. The organization promotes scientific research, ethical considerations, and international cooperation in METI initiatives.

Breakthrough Listen: Breakthrough Listen is a scientific program dedicated to searching for extraterrestrial intelligence by surveying the sky for radio signals and optical transmissions. The program's goals include detecting potential signals from advanced civilizations and investigating anomalous phenomena in the cosmos.

Communication Protocols: Developing communication protocols and guidelines for interstellar messaging is an ongoing effort within the scientific community. These protocols address issues such as message encoding, signal modulation, frequency allocation, transmission duration, message redundancy, and coordination with international stakeholders.

The History of Received Signals

The history of signals received from space is a fascinating journey that spans several decades of scientific exploration and technological advancements. While many signals have been detected and studied, the search for definitive evidence of extraterrestrial intelligence remains ongoing. Here is an overview of key milestones and notable signals received from space:

Radio Astronomy and Early Discoveries

- 1932: Karl Jansky, an American physicist, detected radio waves from the center of our Milky Way galaxy, marking the beginning of radio astronomy.
- 1960s: The discovery of pulsars, rapidly spinning neutron stars emitting regular radio pulses, by astronomers Jocelyn Bell Burnell and Antony Hewish brought attention to the celestial phenomena that emit detectable radio signals.

The Search for Extraterrestrial Intelligence (SETI)

- 1960s-1970s: Interest in the search for extraterrestrial intelligence (SETI) grew, leading to initiatives such as Project Ozma (1960) by astronomer Frank Drake, which scanned the skies for potential signals from alien civilizations.
- 1977: The "Wow! Signal," a strong radio signal detected by the

Big Ear radio telescope at Ohio State University, remains one of the most famous and mysterious signals received. It lasted for 72 seconds and has not been conclusively explained as a natural or artificial orig

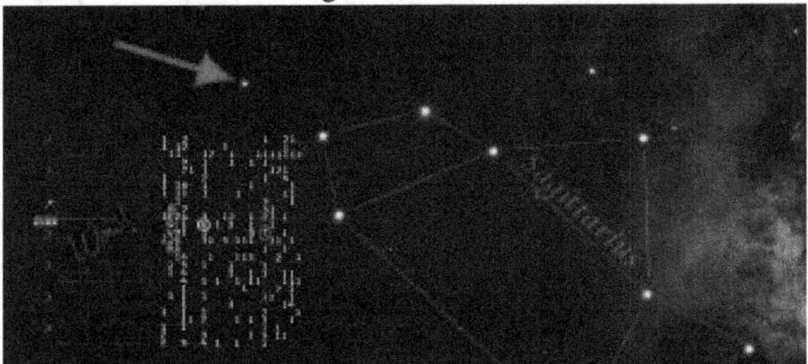

•

Pioneer and Voyager Missions

- 1970s-1980s: NASA's Pioneer and Voyager spacecraft carried plaques and golden records containing messages and information about Earth, intended for potential extraterrestrial encounters. These artifacts were not signals received but were deliberate attempts to communicate with possible alien civilizations.

Discoveries of Exoplanets

- 1990s-Present: The discovery of exoplanets (planets orbiting stars outside our solar system) has expanded the search for extraterrestrial life. Signals such as exoplanet transits (dips in starlight as planets pass in front of their stars) and spectroscopic data revealing atmospheric compositions provide indirect evidence of exoplanets but not direct communication signals.

SETI@home and Radio Telescope Surveys

- 1999: SETI@home, a distributed computing project, launched, allowing volunteers to contribute processing power to analyze radio signals for signs of extraterrestrial intelligence.
- 2000s-Present: Radio telescope surveys, including the Allen Telescope Array (ATA) and the Green Bank Telescope (GBT), continue to scan the skies for potential signals from space. While many interesting signals have been detected, none have been confirmed as originating from an extraterrestrial intelligence.

Breakthrough Listen and METI Initiatives

- -2015: The Breakthrough Listen project, funded by billionaire Yuri Milner, commenced as one of the most comprehensive SETI efforts, using advanced instruments to survey billions of stars and galaxies for potential signals.

Messaging Extraterrestrial Intelligence (METI) Initiatives such as METI International advocate for active communication with extraterrestrial civilizations by sending intentional signals into space, known as Messaging Extraterrestrial Intelligence. These signals are designed to showcase Earth's presence and interest in contact with other intelligent beings. Our advances in technology, including artificial intelligence (AI), machine learning algorithms, and next-generation telescopes, continue to enhance our ability to detect and analyze signals from space. With the development of protocols and guidelines for interstellar communication aims to ensure responsible and ethical engagement with potential extraterrestrial civilizations, addressing issues of mutual understanding, cultural representation, and the consequences of contact. While many signals have been received from space, including natural phenomena, pulsars, and transient events, the search for definitive evidence of extraterrestrial intelligence remains one of the most intriguing and enduring quests in science and exploration. Each signal received adds to our understanding of the universe and fuels our curiosity about the possibility of life beyond Earth.

Chapter 6

Extraterrestrials Chose You

The topic of alleged extraterrestrial implants and their purported use for tracking humans is a subject of much debate, speculation, and controversy within the ufology community. While there is a lack of scientific consensus and empirical evidence supporting the existence of such implants, various claims and anecdotes have been put forth by individuals who believe they have been implanted by extraterrestrial beings. In this discussion, we will explore the concept of extraterrestrial implants, the reported experiences of those who claim to have them, the theories surrounding their purpose, and the challenges in verifying these claims. Did extraterrestrials choose you? From the earliest days recorded events of extraterrestrial implants since March 1957. On a popular radio show named Long John Nebel in 1957 he had a guest Ufologist on the show named John Robinson who claimed to have had an interview with a neighbor who claimed to be abducted in 1938. He said "They placed small transmitters or a headphone device behind my ears to communicate". Also within the timeline in Massachusetts a resident Betty Andreasson claimed that aliens had implanted a device in her nose during her alien abduction in 1967. In these cases all victims recieved some backlash from the public. But hidden in the background after these events strange illnesses were occuring during and weeks after encounters.

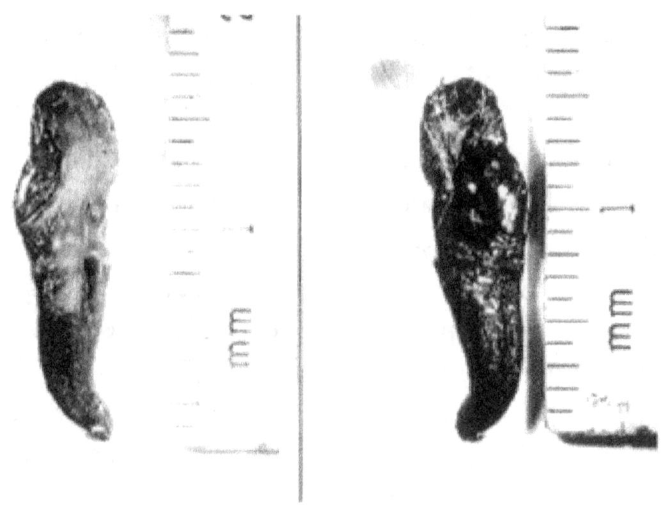

Introduction to Extraterrestrial Implants

Extraterrestrial implants are objects or devices alleged to be placed inside the bodies of humans by non-human intelligence, commonly associated with UFO abduction experiences. These implants are often described as small, metallic, or biotechnological in nature, with properties that make them difficult to detect or remove using conventional medical techniques. Reports of extraterrestrial implants have gained attention through accounts from individuals who claim to have undergone UFO abductions or contact experiences. These individuals often describe intrusive medical procedures during their encounters, including implantation of devices in various parts of their bodies, such as the brain, nasal cavity, extremities, or soft tissues. Dr. Roger Leir was a prominent figure in the field of ufology and alien abduction research, known for his investigations into alleged extraterrestrial implants. His work and findings have sparked both interest and controversy within the UFO community and the broader scientific community. Let's delve into the alien history of Dr. Roger Leir, his research, contributions, and the debates surrounding his work.

Early Life and Background

Dr. Roger Leir was born on October 20, 1934, in California, USA. He pursued a career in podiatric medicine and surgery, becoming a respected podiatrist with a successful practice in Southern California. Leir's interest in UFOs and extraterrestrial phenomena was sparked during his childhood, leading him to explore these topics alongside his medical profession. Dr. Leir gained widespread attention and recognition for his research into alleged extraterrestrial implants. These implants were purportedly objects or devices placed inside the bodies of individuals who claimed to have had UFO abduction experiences. Leir's investigations focused on studying and surgically removing these alleged implants for analysis. Over the years, Dr. Leir performed a series of surgical procedures to extract suspected alien implants from patients who reported anomalous experiences and physical symptoms associated with these implants. His methods involved careful extraction, preservation, and analysis of the retrieved objects. Leir claimed that the extracted implants exhibited unusual properties, such as non-terrestrial materials, advanced nanotechnology, biological integration with surrounding tissues, and electromagnetic characteristics. He argued that these features supported the extraterrestrial origin and purpose of the implants, suggesting they were part of monitoring, tracking, or communication devices used by alien beings.

Controversies and Criticisms

While Dr. Leir garnered support from some ufologists and abduction researchers, his work also faced skepticism, criticism, and controversy within the scientific and medical communities. Critics pointed out several issues with Leir's research, including:

Lack of Independent Verification: Independent verification and peer-reviewed studies of Leir's findings were limited, leading to questions about the scientific rigor and reproducibility of his claims.

Methodological Concerns: Some critics raised concerns about the surgical procedures, chain of custody, and documentation protocols used in implant extraction and analysis, citing the need for standardized protocols and ethical considerations.

Alternative Explanations: Skeptics suggested alternative explanations for the retrieved objects, such as mundane foreign bodies, surgical artifacts, natural anomalies, or misinterpretations of medical imaging.

Despite the controversies and debates surrounding his past work, Dr. Roger Leir's contributions to the field of ufology and alien abduction research left a lasting impact. He authored several books, appeared in documentaries and media interviews, and engaged in public outreach to discuss his findings and theories. Leir's research sparked ongoing discussions about the nature of UFO phenomena, the possibility of extraterrestrial visitation, and the exploration of anomalous experiences reported by individuals claiming contact with alien beings. While his methods and conclusions remain contentious, his work continues to be referenced and debated within the broader context of UFO studies and abduction phenomena. Dr. Roger Leir's exploration of alleged extraterrestrial implants represents a significant chapter in the history of ufology and alien encounter investigations. His efforts to bridge the fields of medicine and ufology, while controversial, contributed to the ongoing dialogue about the nature of UFO phenomena and the search for potential evidence of extraterrestrial contact.

Reported Experiences and Anecdotal Evidence

Those who report having extraterrestrial implants often describe similar patterns and experiences, including:

- Abduction Scenarios: Many claimants recount experiences of being taken aboard UFOs or encountering non-human entities, often described as "greys" or other humanoid beings, who perform medical procedures on them.
- Memory Loss and Missing Time: Abductees frequently report

memory gaps or missing time surrounding their experiences, leading them to suspect that they were subjected to procedures involving implantation.

- Physical Symptoms: Some individuals claim to experience physical symptoms or anomalies associated with the presence of implants, such as localized pain, unusual sensations, electromagnetic interference with electronic devices, or anomalous readings on medical scans.
- Retrieval Attempts: A subset of claimants seek medical attention or undergo procedures to locate and remove suspected implants. However, these attempts are often met with challenges, as conventional medical imaging techniques may fail to detect or locate the alleged devices.

While these accounts provide anecdotal evidence and personal testimonies, the scientific and medical communities generally remain skeptical due to the lack of verifiable physical evidence and the potential for confounding factors such as psychological conditions, false memories, or misinterpretation of experiences.

Theories on the Purpose of Extraterrestrial Implants

Numerous theories have been proposed regarding the purpose and function of extraterrestrial implants, based on the experiences and beliefs of those who claim to have them. One commonly cited theory is that extraterrestrial implants serve as tracking or monitoring devices, allowing alien beings to monitor the movements, physiological data, and experiences of abductees or contactees. Some speculate that implants are used for biological experimentation or genetic sampling, enabling extraterrestrial civilizations to study human physiology, DNA, or cognitive processes. Another hypothesis suggests that implants may facilitate telepathic communication or exert control over individuals' thoughts, emotions, or behavior, although evidence supporting such claims remains elusive. In the context of alleged hybridization programs, implants are theorized to play a role in genetic manipulation, or reproductive procedures aimed at creating hybrid offspring between humans and extraterrestrial beings. It's important to note that these theories are speculative and lack empirical validation, leading to debates within the ufology community and skepticism from mainstream scientific and medical circles.

Challenges in Verification and Investigation

Verifying the existence and function of extraterrestrial implants poses significant challenges due to several factors:

Detection Methods: Conventional medical imaging techniques such as X-rays, CT scans, and MRI scans may not reliably detect small or non-metallic objects embedded in soft tissues or neural pathways, making it difficult to confirm the presence of implants.

Biological Compatibility: The materials and composition of alleged implants are often described as biocompatible or advanced, making them resistant to rejection or detection by the immune system, further complicating attempts at identification.

Chain of Custody: Maintaining a chain of custody and rigorous scientific protocols is crucial in investigating potential implants, as contamination, tampering, or misinterpretation of findings can compromise the credibility of any purported discoveries.

Subjective Experiences: The subjective nature of abduction experiences, memory recall, and psychological factors can introduce biases, inconsistencies, and challenges in separating genuine phenomena from confabulations or hoaxes.

Chapter 7

The CE5 Chosen

Through my years of research of extraterrestrial contact and manifestations, most of the answers were looking for is "Why"? But truly within the realm of CE Starting from 1-5, we should be asking "Who"? I'm a true believer that the experiencer is chosen. Do the entities within our dimension or travelers beyond the universe choose who they contact and why? Possibly through the DNA of your ancestors who were in close connection with these beings in their timeline. You may have that certain structure or strain they're looking for as a contactee or experiencer. The frequencies that we use to contact or wave signal are the beacon as we explained earlier in the segments. The human body is a complex and dynamic system that resonates with a wide range of frequencies, from the cellular level to the entire organism. These frequencies play essential roles in biological processes, physiological functions, and overall well-being. In this exploration, we will delve into the frequencies that resonate within the human body, their sources, effects, and significance in various contexts. The electromagnetic spectrum encompasses a vast range of frequencies, from extremely low frequencies (ELF) to radio waves, microwaves, infrared, visible light, ultraviolet, X-rays, and gamma rays. While the human body interacts with various frequencies across this spectrum, certain ranges are particularly relevant to biological systems and resonance phenomena. The sound frequencies we use during a CE5 session may be what attracts the extraterrestrial to our celestial body. Here are some examples of the frequencies.

Sound and Vibration Therapy

Sound and vibration therapy harness the resonant properties of the human body to promote relaxation, stress reduction, healing, and balance. Key aspects of sound and vibration therapy include:

◇ **Frequency Healing:** Specific frequencies, such as Solfeggio frequencies (e.g., 528 Hz for DNA repair) and Schumann resonance frequencies (approximately 7.83 Hz), are used in sound healing practices to resonate with cellular structures and promote harmonization.

◇ **Sound Therapy:** Sound therapy modalities, such as Tibetan singing bowls, crystal bowls, tuning forks, and sound baths, deliver therapeutic vibrations that resonate with the body's tissues, organs, and energy centers (chakras), promoting relaxation and energetic balance.

◇ **Vibration Platforms:** Vibrational platforms and devices, including whole-body vibration (WBV) machines, deliver mechanical vibrations to the body, stimulating muscles, bones, and circulation. WBV is used in physical therapy, fitness training, and wellness programs.

There is only one way to contact through the consciousness and that's through the brain and the frequencies involved in the process. Here are the possibilities of contact as we function in daily human lives.

Brainwaves and Neural Resonance

Brainwaves, also known as neural oscillations, are rhythmic electrical patterns generated by synchronized neural activity in the brain. These brainwave frequencies correspond to different states of consciousness, cognitive functions, and emotional states:

Delta Waves (0.5-4 Hz): Delta waves are associated with deep sleep, restorative processes, healing, and regeneration. They play a crucial role in the restorative functions of the body, including cellular repair and growth.

Theta Waves (4-8 Hz): Theta waves are linked to relaxation, meditation, creativity, intuition, and subconscious processing. They are prevalent during deep meditation, visualization, and dream states.

Alpha Waves (8-12 Hz): Alpha waves are present in relaxed wakefulness, calmness, focus, and light meditation. They are associated with a state of relaxed alertness and enhanced creativity.

Beta Waves (12-30 Hz): Beta waves are prominent during active wakefulness, concentration, cognitive tasks, and problem-solving. They reflect focused attention, mental engagement, and alertness.

Gamma Waves (30-100 Hz): Gamma waves are associated with higher cognitive functions, perception, learning, and information processing. They play a role in integrating sensory information and synchronizing brain regions. Neural resonance occurs when external stimuli, such as auditory frequencies or visual patterns, synchronize with specific brainwave frequencies, leading to entrainment and alignment of neural activity. This phenomenon is utilized in techniques like brainwave entrainment, binaural beats, and neurofeedback for cognitive enhancement, relaxation, and mental well-being.

The DNA Complex

DNA is and can be very complex with extraterrestrial contact. Thought the Ufologists research since the 1940s through today, the molecular compounds that were most noticed with abductees was a high percentage of *RH Negative* blood lines. Rh-negative blood type refers to the absence of the Rh antigen, also known as the Rhesus factor, on the surface of red blood cells. Individuals with Rh-negative blood lack this specific antigen, while those with Rh-positive blood have the antigen present. The Rh factor is one of several blood group systems, with the ABO system being the most well-known. The distribution of Rh-negative blood varies among populations, with approximately 15% to 20% of people worldwide having Rh-negative blood. The prevalence of Rh-negative blood differs among ethnic groups and regions, with higher frequencies found in certain populations, such as Europeans of Basque descent. Some individuals who claim to have experienced UFO encounters, alien abductions, or contact experiences have suggested a potential correlation between Rh-negative blood type and these phenomena. Anecdotal reports and testimonials have circulated, suggesting that a disproportionately high number of abductees have Rh-negative blood. Within the realm of ufology and abduction research, many accounts and testimonials from individuals who claim to have experienced extraterrestrial abductions sometimes mention perceived demographic patterns, including ethnicity. These accounts are often based on self-reported data and subjective interpretations.

Some human observations regarding ethnic data of alleged abductees include:

Diversity of Experiences: Some reports suggest that individuals from diverse ethnic backgrounds, cultures, and regions have reported experiences of UFO sightings, contact experiences, and abduction narratives.

Commonality of Experiences: Despite cultural and ethnic differences, some abductees describe similar patterns of encounters, including being taken aboard UFOs, undergoing medical examinations, and interacting with non-human entities.

Regional Variations: Most abductees narratives may highlight variations in abduction experiences based on geographic regions, cultural contexts, and prevailing beliefs about extraterrestrial phenomena.

The data is just not there on certain cultural ethnic abduction in this phenomenon. The more that gets reported through organizations such as A.S.D.P and MUFON the more we can see the incoming data and get accurate numbers in such cases. Such cases as the Betty and Barney Hill abduction in 1961 of an interracial couple from Portsmouth, New Hampshire, who were abducted and were taken aboard a spacecraft by extraterrestrials. These aliens showed them their home planet Zeta Reticuli. In which later was discovered through memories of Betty Hill as she slowly recovered from memory loss after the encounter. Later Betty drew a picture of the constellation Zeta that she knew nothing about. Another reported account was in Brazil on October 15,1957. A Brazilian farmer named Antonio Vias-Boas reported in to having an alien abduction story as he was farming a field in the outskirts of town. He claimed to be abducted after seeing a red ball of light come down from the sky after his tractor died, and a UFO craft landed in front of him. A 5 foot all humanoid in a grey jumpsuit and a helmet took him onto the ship and did experiments on him. He stated they made him perform sexual acts with a female extraterrestrial. They gave him a tour of the ship and then escorted him off. Leaving him standing on the road not realizing that 4 hours had passed. So, these 2 cases had similar results but had a different process and end result. An open-minded inquiry should be tempered with scientific scrutiny and rational analysis. Human narratives may mention ethnic data in the context of extraterrestrial abductions, and these claims lack validation and scientific credibility. Cultural interpretations, belief systems, psychological factors, and societal influences contribute to the perception of ethnic patterns in abduction experiences, but caution is warranted in drawing definitive conclusions without evidence. Ethical communication, cultural sensitivity, and critical thinking are essential in navigating discussions about ethnic data and abduction narratives within the broader context of ufology and paranormal research. Personal belief systems, religious backgrounds, and spiritual frameworks can play a role in how individuals interpret and integrate experiences of alleged abductions into their worldview. Beliefs about the nature of reality, consciousness, and the universe can impact the narratives surrounding UFO encounters. To emphasize the belief that ethnic cultures play a role in extraterrestrial abductions is inconclusive. The data has not been fully collected in these circumstances to prove or disprove this theory.

Chapter 8
Meditation of a CE5

Meditation before engaging in a Close Encounter of the Fifth Kind (CE5) experience can be a powerful practice that enhances consciousness, inner peace, and receptivity to subtle energies and higher states of awareness. CE5, coined by Dr. Steven Greer, refers to human-initiated contact with extraterrestrial civilizations through meditation, consciousness expansion, and intentional communication. Incorporating meditation into the preparation for a CE5 experience can facilitate a deeper connection with oneself, the environment, and the potential for contact with non-human intelligence. In this discussion, we will explore the benefits of meditation before a CE5, different meditation techniques to consider, and how to integrate mindfulness practices into the CE5 preparation process.

Understanding the Role of Meditation in CE5

Meditation is a practice of focused attention, mindfulness, and inner exploration that has been used for centuries in various spiritual traditions, contemplative practices, and holistic wellness approaches. In the context of CE5 preparation, meditation serves several purposes.

Centering and Grounding Meditation helps individuals cultivate a sense of inner calm, balance, and presence, which is essential for maintaining clarity, focus, and stability during the CE5 experience.

Heightened Awareness: By quieting the mind and tuning into the present moment, meditation enhances perceptual sensitivity, intuition, and receptivity to subtle energies, including those associated with potential extraterrestrial contact.

Alignment with Intentions: Intention setting is a key aspect of CE5 protocols. Meditation allows participants to clarify their intentions, cultivate positive energy, and align their consciousness with the desire for peaceful, benevolent contact with extraterrestrial beings.

Openness to Transcendent States: Some meditation practices facilitate altered states of consciousness, expanded awareness, and states of deep relaxation, which can create conducive conditions for transcendent experiences and interdimensional communication.

Meditation Techniques for CE5 Preparation

Various meditation techniques can be incorporated into the preparation for a CE5 experience. These techniques can be adapted based on individual preferences, experience levels, and spiritual inclinations. Here are some meditation approaches to consider while beginning a CE5 session.

Breath Awareness Meditation: This classic meditation technique involves focusing on the breath, observing its natural rhythm, and gently redirecting attention back to the breath whenever the mind wanders. Breath awareness promotes relaxation, presence, and mindfulness.

Guided Visualization: Guided meditations utilize spoken instructions or recordings to lead participants through visualizations, imagery, and mental journeys. A guided visualization tailored to CE5 preparation may include imagery of cosmic connections, starlight, and harmonious contact with extraterrestrial beings.

Loving-Kindness (Metta) Meditation: Metta meditation involves cultivating feelings of love, compassion, and goodwill toward oneself, others, and all sentient beings. Practicing loving-kindness fosters a heart-centered approach, unity consciousness, and positive energy, which can be beneficial before engaging in CE5.

Chakra Meditation: Chakra meditation focuses on balancing and activating the body's energy centers (chakras) through visualization, breathwork, and awareness. Aligning the chakras can enhance energetic coherence, vitality, and receptivity to higher frequencies during the CE5 experience.

Sound Healing Meditation: Sound-based meditations incorporate music, chants, mantras, or sound frequencies to induce relaxation, entrainment, and vibrational resonance. Sound healing can harmonize body, mind, and spirit, creating an optimal state for consciousness exploration and contact intentions.

Silent Mindfulness Meditation: Silent meditation involves sitting quietly, observing thoughts without attachment, and cultivating a nonjudgmental awareness of present-moment experiences. Silent mindfulness meditation promotes inner stillness, insight, and the integration of heightened awareness into daily life.

Steps for Meditation Before a CE5 Experience

Here are steps to guide your meditation practice before engaging in a CE5 experience. Set Intentions and begin by setting clear and positive intentions for your meditation and the upcoming CE5 experience. Focus on intentions of peace, harmony, connection, and open-hearted communication with benevolent extraterrestrial beings.

Create a Sacred Space: Find a quiet, comfortable space where you won't be disturbed. Consider creating a sacred atmosphere with candles, incense, crystals, or meaningful objects that evoke a sense of spiritual presence and reverence.

Assume a Comfortable Posture: Sit or lie down in a comfortable position that allows you to relax yet remain alert and attentive. Close your eyes gently or maintain a soft gaze, whichever feels most natural for you.

Focus on the Breath: Begin by bringing your awareness to your breath. Notice the sensations of inhaling and exhaling, the rise and fall of your abdomen or chest, and the flow of air through your nostrils. Allow your breath to anchor you in the present moment.

Body Scan and Relaxation: Gradually scan your body from head to toe, releasing any tension, tightness, or stress you may be holding. Soften your muscles, relax your facial expressions, and let go of any mental or physical distractions.

Choose a Meditation Technique and select a meditation technique that resonates with you, such as breath awareness, guided visualization, loving-kindness, chakra meditation, or sound healing. Follow the instructions for your chosen technique, allowing yourself to immerse fully in the experience.

Visualize Cosmic Connections: If practicing a guided visualization, imagine yourself surrounded by cosmic light, interconnected with the universe, and open to receiving higher frequencies of consciousness. Visualize harmonious contact with extraterrestrial beings, characterized by peace, respect, and mutual understanding.

Cultivate Positive Emotions: During your meditation, cultivate feelings of love, gratitude, compassion, and joy. Imagine sending waves of positive energy into the cosmos, radiating goodwill toward all beings and inviting benevolent contact with extraterrestrial intelligences.

Stay Open and Receptive: Remain open and receptive to whatever experiences, insights, or sensations arise during your meditation. Trust your intuition, inner guidance, and sense of connection with the greater cosmic consciousness. End with gratitude and conclude your meditation with a moment of gratitude for the opportunity to connect with higher realms of consciousness, explore the mysteries of the universe, and participate in the CE5 experience. Express gratitude for the support of benevolent beings and the guidance of universal wisdom.

Integration and Post-Meditation Reflection

After completing your meditation practice, take a few moments to integrate the experience and reflect on any insights, feelings, or inspirations that arose. Consider journaling about your meditation experience, recording any messages or impressions received, and noting any shifts in consciousness or awareness. Engage in self-care activities, such as gentle movement, hydration, nourishing food, and rest, to support your well-being and integration of the meditation's effects. Maintain a sense of openness, curiosity, and receptivity as you approach the CE5 experience with a heightened awareness and attunement to cosmic energies. Meditation deepens self-awareness, expands consciousness, and fosters a sense of interconnectedness with the cosmos, preparing you for heightened states of perception and awareness during the CE5 encounter. The most important factor to achieve during and after a CE5 is your Inner Peace and Calm Meditation cultivates inner peace, emotional balance, and calmness, reducing stress, anxiety, and mental chatter that may interfere with your CE5 experience.

Digital Data Reference

Record your experience frence with a written log or digital device such as cell phone or handheld recorder. Plack back recordings with headphones or large Bluetooth speakers and listen for voices or certain tones with the CE5 meditation process was in session. Check SD cards in digital cameras using software to see any disturbances of a digital signal that you may analyze yourself or someone who is more experienced in the digital field analysis of spectrum. Remember the consciousness and subconscious during your session was a universal and a perceptive contact willingly. Wavelengths can be inverted into digital signals through contact manipulated by the extraterrestrial for you to understand. The communication may not be verbal but will be through universal and point to point contact. You may not always be successful in every session of contact. But be aware once you knock on the door someone or something will eventually answer. Check your video footage as well for conscious contact using frequencies listed earlier. Playback in real-time the video footage as the session was being documented. Look for digital interference during the CE5 session. Look for glitches or smoke like iridescent shapes in the data recorded. You may be able to zoom in on still and get an image of extraterrestrial entity as it appears on the video data. Record time and date on a written log of entity caught in data. This information is useful for future CE5 references and data scaling.

Chapter 9
A.L.I.E.N.S- (EBE)

Within the realm of ufology, science fiction, and speculative theories, various alleged extraterrestrial species have been described based on reported encounters, abduction narratives, and conspiracy theories. It's important to note that these descriptions are speculative and through some witnessed accounts recorded over the centuries of time.

Here, we will explore some of the commonly mentioned extraterrestrial species in popular culture and ufology, along with the characteristics attributed to them.

Greys (Zeta Reticulans)

One of the most well-known and iconic extraterrestrial species in ufology is the Greys, also referred to as Zeta Reticulans. They are typically described as humanoid beings with grey skin, large heads, almond-shaped black eyes, small noses, and mouths. Greys are often depicted as slender, with long limbs and fingers. They are commonly associated with abduction narratives and purported to be involved in genetic experimentation, hybridization programs, and telepathic communication.

Reptilians (Reptoids)

Reptilians, also known as Reptoids, are a controversial and often sensationalized species in ufology and conspiracy theories. They are described as humanoid or reptilian beings with scaly skin, reptile-like features (such as slit eyes and elongated faces), and sometimes tails. Reptilians are often portrayed as highly intelligent, technologically advanced, and possessing psychic abilities. Some conspiracy theories suggest that Reptilians have infiltrated human society and hold positions of power. Some others believe that they live beneath the earth in caves or underground facilities.

Nordics (Pleiadians)

Nordic aliens, also known as Pleiadians or Tall Whites, are described as humanoid beings with fair skin, blonde hair, and blue eyes. They are often depicted as tall, elegant, and possessing a benevolent and spiritually advanced demeanor. Nordic aliens are associated with messages of peace, enlightenment, and cosmic consciousness in some urological and New Age circles. They are purported to have telepathic abilities and to communicate messages of spiritual evolution and unity.

Insectoids (Mantids, Ant-like Beings)

Insectoid extraterrestrial species, such as Mantids and Ant-like Beings, are described as insect-like or arthropod-like creatures with multiple limbs, antennae, and segmented bodies. Mantids, in particular, are often depicted as tall and slender beings with insect-like features, including large eyes and elongated limbs. Some abduction accounts and contactee narratives mention interactions with insectoid beings, who are sometimes described as overseeing or guiding human evolution.

Humanoids (Andromedans, Arcturians)

Humanoid extraterrestrial species encompass a wide range of descriptions, including beings from star systems such as Andromeda and Arcturus. These humanoid aliens are often depicted as resembling humans in appearance but may have subtle differences, such as larger eyes, different skin tones, or unique physical characteristics. They are associated with advanced technologies, spiritual wisdom, and a peaceful approach to interstellar relations in ufological lore.

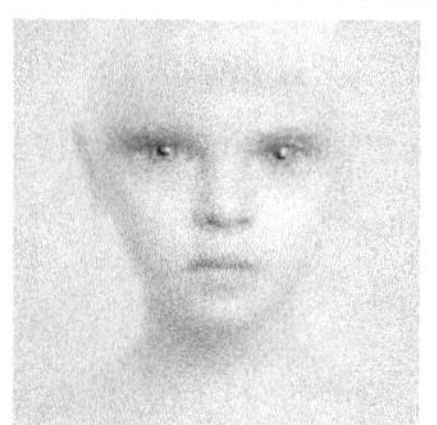

Tall Whites (Tall Greys)

Tall Whites, also known as Tall Greys or Tall Whites Greys, are described as taller versions of the classic Grey aliens. They are depicted as having similar features to Greys but with elongated bodies, taller stature, and a more commanding presence. Tall Whites are sometimes associated with military-related abduction narratives and alleged interactions with government agencies.

Human-Analogues (Hybrids, Human-Like Species)

Some alleged extraterrestrial species are described as human-analogues or humanoid hybrids, blending characteristics of humans and other extraterrestrial beings. These hybrids may have features such as mixed physical traits, telepathic abilities, and a connection to both human and alien civilizations. Hybridization narratives often involve themes of genetic experimentation, interbreeding, and the creation of hybrid offspring.

Energy Beings (Light Beings, Higher-Dimensional Entities)

Beyond physical forms, some reported extraterrestrial encounters involve encounters with energy beings, light beings, or entities from higher dimensions. These beings are described as non-corporeal, luminous, and existing beyond conventional spacetime constraints. They are associated with spiritual teachings, consciousness expansion, and multidimensional realities in ufology and metaphysical discourses.

Gray skin/Human Hybrids

Some accounts describe hybrids that possess features of both Greys and humans. These hybrids may have grey skin but more human-like facial features, such as noses and mouths. They are often depicted as intermediaries or ambassadors between human and Grey civilizations, facilitating communication and interaction between the two species.

Artificial Intelligence (AI) Entities

In speculative ufology and futurism, there are theories about extraterrestrial civilizations that have evolved into artificial intelligence entities. These AI entities may exist as advanced consciousnesses within technological constructs or interstellar networks. They are associated with transhumanist themes, digital consciousness, and the integration of biological and technological intelligence.

Interdimensional Entities (Shadow Beings, Interdimensional Travelers)

Some reported encounters with extraterrestrial beings involve entities that transcend conventional space-time boundaries and exist across dimensions. These interdimensional entities may appear as shadow beings, light orbs, or shape-shifting forms. They are associated with metaphysical concepts of parallel realities, alternate dimensions, and quantum consciousness.

Aquatic/Marine Species (Water-based Beings)

In speculative ufology, there are theories about extraterrestrial species that have aquatic or marine origins, residing in oceans or water-rich environments on other planets. These water-based beings may have amphibious characteristics, aquatic physiology, and advanced technologies suited for underwater habitats. They are associated with aquatic civilizations, underwater UFO sightings, and oceanic mysteries.

The importance of emphasizing that the descriptions and classifications of alleged extraterrestrial species are speculative and based on reports, abduction narratives, contactee accounts, and conspiracy theories.

Chapter 10
 The Final Contact Balance

After a Close Encounter of the Fifth Kind (CE5) contact with a positive result, you may experience a range of transformative effects that can profoundly impact their life, consciousness, and worldview. CE5, as defined by Dr. Steven Greer, involves human-initiated contact with extraterrestrial civilizations through meditation, consciousness expansion, and intentional communication. A positive result in a CE5 encounter typically refers to peaceful, benevolent interactions with non-human intelligence.

One of the most profound effects of a positive CE5 contact is an expansion of consciousness and awareness. Contactees often report heightened states of perception, increased intuition, and a deeper connection to universal consciousness. They may experience a sense of interconnectedness with all life forms and a greater understanding of the cosmos and their place within it.

Spiritual Awakening and Transformation -Positive CE5 contacts can trigger spiritual awakening and transformation in individuals. Contactees may undergo profound shifts in their beliefs, values, and spiritual practices. They may explore concepts such as unity consciousness, higher-dimensional realities, and the interconnectedness of all existence. This spiritual awakening can lead to a deeper sense of purpose, inner peace, and alignment with universal principles.

Healing and Energy Balancing - Some contactees report experiencing healing and energy balancing during or after a positive CE5 encounter. This healing may occur on physical, emotional, mental, or spiritual levels. Contact with advanced extraterrestrial civilizations is believed to involve energy exchanges and vibrational healing that can promote wellness, vitality, and holistic well-being.

Expanded Psychic and Intuitive Abilities - Positive CE5 contacts can activate or enhance psychic and intuitive abilities in individuals. Contactees may develop telepathic communication skills, clairvoyance, precognition, empathic sensitivity, and other extrasensory perceptions.

These abilities can facilitate deeper connections with others, intuitive insights, and expanded understanding of multidimensional realities.

Heightened Emotional States - Contactees often describe experiencing heightened emotional states during and after a positive CE5 contact. These emotions may include joy, awe, gratitude, love, and a profound sense of wonder. Some individuals report transpersonal experiences, such as feelings of cosmic unity, oneness with the universe, and encounters with beings of light or higher consciousness.

Integration of Cosmic Wisdom and Insights - After a positive CE5 contact, individuals may integrate cosmic wisdom, insights, and teachings into their daily lives. They may gain new perspectives on planetary issues, environmental stewardship, peaceful coexistence, and the evolution of consciousness. Contactees often become advocates for unity, cooperation, and harmony on Earth and beyond.

Enhanced Creativity and Inspiration - Positive CE5 contacts can inspire creativity, innovation, and artistic expression in individuals. Contactees may channel cosmic inspiration into creative projects, music, art, writing, and visionary endeavors. They may tap into universal energies and collective consciousness to create positive impact and transformative works.

Deepened Connection to Nature and Gaia Consciousness - Many contactees report a deepened connection to nature and Gaia consciousness after a positive CE5 encounter. They may feel a stronger bond with the Earth, its ecosystems, and natural elements. This connection can lead to eco-consciousness, environmental activism, and a commitment to preserving and honoring the planet.

While a positive CE5 contact can be a profound and transformative experience, it may also present integration challenges and growth opportunities. Contactees may need time to process and integrate their experiences, adjust to expanded awareness, and navigate changes in their belief systems and relationships. Supportive communities, spiritual practices, and self-care strategies can aid in the integration process. Many contactees feel compelled to share their experiences with others and connect with like-minded individuals who have also had positive CE5 contacts. They may join CE5 groups, participate in consciousness research, attend gatherings, and contribute to the global dialogue on extraterrestrial contact, consciousness exploration, and the evolution of humanity.

A CE5 encounter can ripple through every aspect of a person's life, fostering growth, connection, and alignment with higher principles of love, unity, and cosmic consciousness.

The final balance is, if you've made contact or not, it's OK. If you feel good about what has happened around you while attempting to do the CE5. The meditation that you do while interpreting and doing the CE5 event Is key. This helps the mind, body and soul. The music and the frequencies soothe your Aura and your chakra. Just think of this as a mind, body and soul exercise for your balance and existence. The information I provided gives you a sense of meditation almost like an astral projection session. The universe is filled with many frequencies and outcomes and inbounds of messages sent and or to be received. Just know that the projection that was attempted and received through all started with you. You are key to the connection of any extraterrestrial or universal contact. Make sure you're at right body sound and mind clear headed with the right intentions. Make sure you always have a safe environment and provide the Extraterrestrial with safe and loving intentions. Assure them that no harm will come to them or to yourself as a C5 is taking place and being recorded. Ask them to allow the data recording and respond to any questions or answers given by you. And make sure while you're meditating to allow nature to resonate with you. By being forthcoming in the right state of open mind the probabilities of contact go up much higher. If you don't succeed with the first couple of CE5 attempts continue to do longer meditations before going forward.

I hope this book helps you Whatever's making contact with you through conscious or subconscious approaches. Just know that I am with you in your endeavor of what you are looking for. I'm still seeking the same answers as you. I'm still asking. Are we alone? Are there extraterrestrials among us? Are they contacting me through the conscious or subconscious? Are they here to help me? Or are they here to harm me?

I believe you can contact benevolent creatures or extraterrestrials through these exercises. Remember, keep your eyes on the skies. Keep your body, mind and soul in a good place. And keep reaching for CE5 contact.